Brethren Composer

The Autobiography of George B. Holsinger

With a Foreword by
Charles Culbertson

Clarion Publishing
Staunton, Virginia

Library of Congress Control Number: 2018937274

ISBN: 978-0-9908190-9-7

Clarion Publishing
Staunton, Virginia
clarionpublishing@gmail.com

George Blackburn Holsinger

Photo by Dean's Studio, Harrisonburg, Virginia

Table of Contents

Images

Foreword

The handwritten autobiography of Brethren composer George B. Holsinger (1857-1908) was acquired in the early 1980s at a Mt. Crawford, Virginia, auction. It lay ignominiously at the bottom of a cardboard box brimming with unrelated paper items that included old billheads, scraps of business correspondence, half-shredded newspapers and stray photographs.

Published here for the first time, Holsinger's unfinished, 1905 autobiography sheds valuable light on the man who composed more than 200 hymns and was the first head of the music department at what would become Bridgewater College in Bridgewater, Virginia.

The Pennsylvania-born Holsinger, who jotted down this sketch of his life for the Brethren Publishing House at Elgin, Illinois, is renowned for his contributions to Christian music, but less is known about his private life. Holsinger died without ever having finished this manuscript or provided any part of it to the Elgin publisher. His autobiography reveals, anecdotally, some of his chief characteristics and closely held beliefs.

Biography

Born May 10, 1857, on his father's farm near Pleasantville, Pennsylvania, George Blackburn Holsinger was the oldest of four sons of Joseph H. and Rebecca Blackburn Holsinger. For a number of years his father was the leader of song for the Church of the Brethren, and George Holsinger grew up surrounded by the beauties of nature, which he appreciated to the end of his days. Because the nearest school was two miles away, he was taught by his mother in his early years and later received eleven years of common-school education.

Holsinger prepared himself for teach in the common schools, but—having a passion for music—drifted into the study of music. In 1875 he attended a singing-school taught by Benjamin Holden Everett (Holsinger mentions him in his autobiography), and then began class-work on his own. He played the organ long before he could read music, and used chords before he knew how to make them other than through his inborn ability.

In 1881 he attended a musical institute where, under W.R. Burnett, he received his first insight into reading music and playing the organ. At this institute he took first prize in song composition and did likewise the year following. Eager for the best, he studied under such men as B.C. Unseld, George F. Root, H.R. Palmer, J.C. Filmore and others. It was this high-grade education that undoubtedly prepared him for the volume and high character of his own work.

When Bridgewater College, then known as Virginia Normal School, was founded in 1882, Holsinger took charge of the music department and remained in the position until 1898. He taught piano, organ, vocal, history of music, harmony and composition. He gave more than 300,000 lessons of instrumental music alone.

In 1898 the Brethren Publishing House offered him the position of musical editor, and from that time until his death in 1908. He prepared songbooks and traveled among churches conducting singing classes and musical institutes.

As a musical composer and editor, Holsinger's life was a marked success. The 200 songs he composed—a number of which his wife wrote the words for—found their way into more than 100 collections.

"While here in the body," noted his wife, "his spirit seemed to take flight when composing, and so some of his best songs anticipated the time when he, with all the redeemed sinners, should get home. Thus it is noted that such as *Rest Over Jordan*, *At the Savior's Right Hand* and *Meet Me There* are songs of this character and will be sung in the years to come.

"Then," she continued, "as though realizing that he was on Earth, he breaks forth with another, *Keeping Step With Jesus*, indicating how closely he sought to associate with his Master. There is a melodious sweetness in his music peculiar to his compositions that reveals the trust and gentleness that were so marked in him."

In the last decade of his life, Holsinger edited and assisted in editing nine different volumes of song and published a book of practical exercises that was helpful to teachers of vocal classes. His best work—or rather the one that served the largest field of usefulness in its day—was the *Hymnal* of the Church of the Brethren, published in 1901. Additionally, he conducted classes in eleven states. Within this period he was leader of song at nearly every Annual Conference.

On October 8, 1875, Holsinger united with the Church of the Brethren, and was known for his piety. On August 9, 1884, he married Sallie A. Kagey of Bridgewater, Virginia. She has been described as poetical, appreciative and devoted, and supplemented her husband's work in every way. When he was away from home for long periods, as he was in his later years, Sallie refused to complain, but instead encouraged Holsinger in his work. Twin boys were born to them, but Clyde Kagey Holsinger alone reached maturity.

As his own words in this book indicate, Holsinger was an easily discouraged man who was also sensitive to an insult or perceived injury. However, he was not retaliatory and never sought revenge on those whom he thought slighted him.

"He was modest almost to a fault, and yet could make himself at home before the largest audiences," wrote his wife. "His hand was always there to help another, and in his chosen profession he had many opportunities and used them. As a teacher he did not think of his ability, but of how to make still better progress with his pupils. He was always ready for his class-work, knew what he was going to do and usually accomplished it."

It was while engaged in teaching a vocal class at Astoria, Illinois, that Holsinger fell ill with pneumonia, and died on November 22, 1908, before either his wife or son could reach him. His last words after his anointing, and just before his death, were, "I have a narrow passage to go through. If it were not for my wife and poor child, I would be ready to lie down and go to my Father."

He was buried in Oak Lawn Cemetery in Bridgewater, Virginia.

Editorial Note

Holsinger's unfinished, 1905 manuscript provided a number of editing challenges, the most significant of which was the "Chapter of Accidents." In the manuscript, Holsinger broke off the narrative of his life in mid sentence to begin the "Chapter of Accidents," and then continued his narrative in an unreleated spot. It is unclear exactly where he wanted the "Chapter of Accidents" to appear, so a bit of editorial divination was required; we can only hope that we placed it in an appropriate location.

Other than that, and the repositioning of a single sentence (in this instance, in the margin, Holsinger *did* indicate where the sentence should go), Holsinger's manuscript appears as he wrote it. His misspellings are indicated by the use of "sic" and the portions he left blank for want of a date or phrase are also indicated. He began writing in the third person and by the fourth sentence was writing in the first person; we have not corrected that, as well. There are also discrepancies in some of the dates he provides (he gives his mother's date of death as both 1881 and 1882). All this we have let stand.

Charles Culbertson
Editor

Brethren Composer

Preface

This autobiography, begun at the request of Bro. Galen B. Royer, for the Brethren Publishing House at Elgin, was begun April 4th, 1905, at 1:30 p.m. in the home of Bro. D.S. Wampler of Timberville, Va. I came to Timberville the day before to begin a vocal class to last about two weeks.

This is intended only as a rough sketch or outline of events as they come to my mind and some day may be revised, classified and rewritten.

The sketch may be intended as a little remembrance to future generations should my life prove interesting enough to interest others.

This may never happen, for few men attain enough eminence in the works they do to be remembered beyond their immediate relatives and friends. "Fame, thou art a bubble."

Autobiography of

George B. Holsinger, born Sunday morning, May 10, 1857 (born about 8 a.m.) in the old log house then owned by his father, and where most of his days were spent. His father and mother never moved about much, but it is thought they moved from the home place to a place called "The Shop" or saw mill, which consisted of a saw mill, a turning shop and carpenter shop. His (G.B.H.'s) earliest recollection is of things that happened while living in this saw mill. I remember that mother cut a hole, or rather two of them, in the plank wall and put glass therein, so that when mother would go to the spring near by, or would go out to milk, brother Joseph B. and I could look out of these "our" windows. My age was certainly not more than between two and three. I also remember how brother J.B. and I would go under this mill to watch the great wheel go round. While living in this mill I once fell from a workbench, striking a block or slick which has left a mark on my right cheek, still visible. The old saying "brought up on a saw mill" is literally true and applies fully to the subject of this sketch.

In 1861 or 1862, father moved from the shop to the "New House" as we always called the house he built in the same hollow. While living here, atho (sic) but very few things can be remembered, there were a few little happenings that to me are worthy of recording. My most vivid recollection is of mother killing a house snake in the platform to the cellar way. Mother was always very brave in killing a snake and no snake was too large or fierce for her to tackle. I can remember her excitement at the idea of a snake being so near her boys and alive in the house. I can remember the water from the spring above the house running through the trough in the basement or cellar, also how the little run in front of the house suddenly rising from heavy rains, backed up in to the cellar and before father and mother could get to the trough the milk was "drowned." This "drowning" of the milk I could not understand, and it impressed my mind in such a way that I could never forget it.

Once on a time, brother J.B. and I were playing about the spring above mentioned when J.B. tumbled in. The water being shallow and he seeming to have his wits turned around and cried as I suppose so that when mother came to the rescue he was standing with his hands on the rock in front of the spring.

Father was building a porch to this "New House" and I have heard him relate how frightened he was when he discovered that a hatchet he had laid on one of the joists was just in the act of slipping off, which had it occurred would have fallen directly on the head of his oldest son, Geo. B. Aside from the above very little else do I remember of the new house. And yet strange as it may seem there seems to have been created within my nature a great love for this home, that has come down through all these years. As evidence I relate that very many times have I found myself weeping in my sleep because we were about to move away from the "New House" or had done so. These dreams have been many times repeated.

In 1863 we moved from the "New House," I think to Grandfather Blackburn's in what we called "Chestnut Ridge," near Schellsburg, Pa. Although I was older by several years I do not remember very many things that happened while living here. We stayed with Grandfather, farming for him only one year.

I can remember how brother Jos. B. and I played under a large white pine tree up against the side of the hill and listened to the whisperings of the wind as it passed through its great branches. I can remember we talked of this mysterious sound and wondered why the wind did not whisper when passing through other trees. Then I have a sleight (sic) recollection of the little one room log house where we lived, of boring with the "big gimlet" some holes in the logs to which I had made some spool wheels to run. Only a very few things else do I rem[em]ber and they are too triv[i]al to mention.

I do rem[em]ber our moving back to the "old home," the log house, in 1864, in which I was born. I remember how when we came to "Dunning Creek," to us the largest creek we had ever seen, and how I wondered how old "Bull," the family dog, would get across. Imagine my astonishment when he plunged into the water and swam across. The old dog lived many years afterwards, dying of old age. A day or two before he died, by some means or other, he lost the hide and hair of his tale (sic), nothing remaining but the bony vertebrae of this member, that had so often wagged a welcome more eloquent that words, to returning members of the family.

Then I also remember of this my parents last flitting, how that after we got home and had unloaded and I could get to my box of playthings, how that I chose the south window in the large rom (sic) for my window, pouring out the playthings on the window cill (sic).

This much for the ancestry of our family. The name being German it was spelled Holtzinger. It is said that persons by our name still live in Germany. I hope that someone will finish a history of the Holsinger family begun by H.R. Holsinger who died only last month, March 13, 1905.

My father was a descendant of Rudolph Holsinger who emigrated to this country Sept. 21, 1731, or just about one hundred years before he was born.

My mother's name was Rebecca Blackburn, second daughter of Joseph Blackburn and Susannah (Wonders) Blackburn. Grandfather was a small man but remarkably quick and agile. Mother partook of his characteristics to a great extent. As a girl she could do the work of a man and almost anything a man could do. She delighted to do work in the harvest field, on the clearing, anywhere in fact that her help was needed. She had a rather remarkable mechanical talent for a woman. She made me the first pair of leather shoes I ever wore, a child's chair which I still possess, and with her own hands turned all the rounds, &c., platting or plaiting the bottom of oak splints, that father had made. This and three cross pieces in the back was all she did not work out of the rough.

Father was a tall man measuring six feet in his "stocking feet" as they used to say. He was tall and slender, and never possessed robust health, after he had a winter's siege of rheumatism in the year 1852 when he was 21 years of age.

He, too, was remarkably endowed along the line of mechanics. It seems to me there was nothing that he could not do, from fixing or repairing the clock to making a wagon. He always kept the clock in repair, made his own shoes and boots, and mended his as well as of all the family, for he owned a kit of shoemakers' tools. He owned blacksmith tools and if the bellows cover wore out he put on new leather, made his own horse shoes and shod his horses, made their harness, put tires on wagons, made the wood work of wheels, running gear &c. and ironed it, made his sled out and out, painting same &c. He made all his cabinet ware such as bookcases, beds, chests, tables, chairs &c ., built houses, barns and any kind of building. He would buy up an old threshing machine, build the woodwork to same, and many a long day did we thrash wheat with machines of his make. After father saw some plasterers plaster the upstairs rooms, he concluded he could do it, too, and he proceeded to plaster the "back" room of the first story.

After we moved back to the old home I somehow do not have a memory of very many special events for some years. I should think that we moved back to the original house in 1860 to 1862. Since I kept no diary, any events spoken of must go without dates. My first attempt at keeping a diary was in 1872 when I was 15 years old.

One event of importance was the tearing down of an old stone kitchen with walls two feet thick and replacing it with a two-story plank building making our house quite roomy. When it was decided to tear it down I recall vividly how bro Joe B. and I mounted the roof and began to knock off the shingles. But it was no small task to remove the stone walls. Finally all was hauled away and the new two-story part was builded (sic), giving us three large rooms downstairs and about five upstairs.

For a long time the second story of the log part was all in one room. The partitioning off of this large room and the plastering thereof was an event not forgotten.

Then came another event. The old log barn had to be torn away and a nice frame barn arose in its stead. This was done in

1872 and my first diary contains an account of this important event.

Having several miles to [go to] school and a mother unusually solicitous as to the welfare of her children, I did not start to school till after I was about eight years old. I do not think father or mother went with us as they should probably have done, but brother J.B. who was a year and a half younger than I was, started together alone. I have been told by those who were there and remember it that we were a wild looking, scared set as we walked or crept up the side aisles to a seat.

I have one little memory vivid and clear of this first year's schooling. Sometime during the year all the pupils were to have a declamation. I knew nothing as to what would be appropriate, and I hardly think father or mother had a clear understanding as to what was wanted.

I looked through my spelling book and McGuffey's reader and finally decided [t]hat a certain little reading exercise would do. I committed it if it required committing.

The day arrived for the speeches. My turn came. I suppose I looked and acted rather scared, and by the time I got on the rostrum and behind the long moon shaped desk I was more ready to hide than to say my speech. I proceeded to <u>hide</u>. All the pupils laughed and no wonder. Then the teacher said I must stand up and say my speech. I stood and said one half of it which was "This boy has a bird." Then all the pupils laughed again. I was ready to do some more <u>hiding</u>, which I proceeded to do. Again the teacher came to the rescue and coaxed me to stand up and finish my speech. The balance of it which ran "Some birds can sing" caused another roar of laughter. How I got away from that dreadful scary place I do not rem[em]ber, but this I do know— that I never became famous as a reciter. I do not know that the first dreadful experience had much to do with my failure along this line. Children of today are saved such humiliating experiences by their teachers' good judgment in the selecting of their speeches.

As a pupil I think I was not such an unruly boy as many, for

I never but once was punished. This punishment was simply that a schoolmate and I was called out by the teacher to sit on the dunce bench. I was not the guilty one but had to stuffer and I suppose the teacher either regretted the punishment meeted (sic) out or smiled at my bauling (sic) on account of it.

However, I can not but think that I was not a little mischievous and that many times merited punishment that I did not receive.

I do not think I was brilliant as a pupil in my studies at all. I was a miserable speller. I always stood at the foot end of the spelling class. I remember now that on one occasion the whole class missed the little word "blood." It had gone around the class more than once. Finally the teacher gave a little hint to a pupil above me that he failed to grasp, but I did, and was a proud boy when I walked in front of that whole class and took my place[at the] head. It was the last word, or at least the teacher made it so, possibly for my encouragement, so that I would get a head mark once.

My lack of skill at spelling stuck to me long after I began to teach school even. At a teacher's examination I remember missing twenty-one words out of twenty-five. However, they were really difficult words. I never learned to spell till I used the method of spelling at a word and then looking it up in the dictionary.

One strange feature about this poor spelling was that it did not bother me in my reading, for I was considered a rather good reader. I am not an adovcate of sending children to school too young, at least if the teaching today were no better than it was when I went to school. I can well remember that I could get along very well in arithmetic till we would get to common fractions, and then I seemed under the teaching of those days to be utterly unable to comprehend what they meant. I would run up against them, then have to turn back and get that far again and again, till finally my comprehension widened and I was able to pass over what had seemed to be an insurmountable obstruction or barrier. Finally I learned to be rather skillful with figures but not a prodegy (sic) by any means.

I should have mentioned in connection with my ability at reading that the first year I attended school I read in the first reader. The second year I was advanced to the second reader and each year I went one reader higher till the fifth reader was reached.

I had shown some talent in learning my letters by seeing the first large letter at the head of the first word of each chapter of the Bible. I think I learned all the capital letters in this way. Then mother took a thin board about a quarter of an inch thick, four inches wide and about eight in length, and on this she pasted a piece of white paper, and from paper and almanacs she cut capital letters and small letters, and from this board I learned my letters. She also made some letters into words and sentences, and from this board book I learned to read. Perhaps the book was too <u>hard</u> which may account for my being such a hard—poor—speller.

I mentioned that mother cut some of the letters from almanacs, and as an evidence of or proof I mention that I happen to possess—thanks to their preserving care of the almanac of the year in which I was born—1857. The cover of this almanac contains the holes of the clipped letters.

My public school days were about ten or eleven years. I should like to give the names of my teachers but this would only be interesting to the writer.

After the accident of the cut in my knee mentioned elsewhere, in 1876 I attended my first Normal school for public school teachers. This was quite an interesting experience to me. My teacher's name was ___ Mehamar. I boarded at Eld. Leonard Furry's pleasant home near New Enterprise, Pa. The school lasted ten weeks, and when it was over I was thought to be able to teach school. Sister Emma A. Miller, daughter of ____ Miller and since wife of Eld. Geo. Replogle, thought I could get a school in her township in Miligan's Cove. But for some reason the directors did not take to me kindly and I failed to get the school she wanted me to have. I think it was the impression that I would succeed better away from home than to undertake to teach a school in my home township. Why this impression I can hardly surmise, but think that as a boy or even a young man of 19 I did not make a very good impression on the directors and patrons. Or I might express it in another way using the words of my dear uncle Eld. C.S. Holsinger, who said that he once thought that "Geo. B. never would amount to very much." I seemed to have little ambition, and yet there was latent talent that only needed to be aroused to change uncle's opinion. I believe I did not give his entire expression above, which was made after I began to show some energy and interest in the line of music.

His expression was, "I never thought George would amount to very much till he began to learn music."

In another chapter I will speak of my musical studies, ambitions and successes.

To resume my story of my school days along with the story of the five years that I taught public school, I should say that I have always considered that the failure to get the school in Milligans Cove was a blessing to me in disguise. I have always thought that had I gotten the school that I should have signally failed with it, which should it have occurred would have so discouraged me that I could not have had the success in teaching school that I can hereafter record.

The school was large, rather hard to manage and had many grown pupils in it. My inexperience and necessary knowledge was surely against my success. Be this as it may I have to record a little incident that occurred the following winter that I regret to this day. As I had failed to get a school and was in my twentieth year it was my privilege to attend public school. The winter before this we had a dear good teacher who had it seems to me taken an unusual interest in me, and who welcomed be hack into her school with delight. For some months all went well till one day in parsing a sentence in grammar my teacher did not apply the rule as was the custom. in order to give the reason for doing a certain thing. The ten weeks in the Normal I suppose had made me somewhat conceity (sic) of my knowledge of grammar and so sprang a dispute that ended in estrangement and I quit her school. I wrote to some grammarians and got opinions that seemed to strengthen my position which the teacher was not willing to accept fearing it would weaken her prestige. Neither of us gave up.

After I got to teaching I saw the unkindness of my procedure and have always regretted the course I took. Many times did I desire to ask the teacher's pardon but before I got to do so typhoid fever had closed her lips in death and the desired opportunity had fled. But I sufferred (sic) many a regret. I felt I had not done as I should. Even tho I may have been right I should have respected my teacher's opinion and especially refrained from hurting the feelings of a young lady who as my teacher was so kind to me and was really more interested in me than I know, as I have evidence to show, but did not then realize it as I did afterwards, and do at this writing after having lately read her letters written to me after the first winter's school. Her name was Jennie Hammaker.

But there seems to have been some providential guidance in these little incidents that only was intended to give me other opportunities, for after I no longer felt like attending the home school, Sister Emma A. Miller wanted me to come and attend the very school she had tried to get for me to teach. She was a most excellent teacher and I learned many things from the books and especially along the line of discipline that was helpful to me in my schools that I afterwards tried to teach.

A Chapter of Accidents

It is said that we never know how to handle any tool well till we cut ourselves with it. If this be so then I should be a thorough mechanic of many trades.

It would seem that I must have had rather more than my share of accidents, and of many kinds.

One of the first occurred when I was a mere lad and I hardly think I have any remembrance of it, and yet as I write I seem to recall it quite vividly. We had a colt some months old and it may have been near a year old, that we all thought much of. Father was starting towards the house from the stable. I was walking close behind him. I saw father place his hand lovingly on the colt's back and stroke it. As I came along I thought I could do so, too, and proceeded to be rather too familiar as the colt thought. The colt turned his heels towards me and began kicking me. I suppose I was too badly scared to run and father hearing an unusual noise turned and saw the colt pinning me against the barn with both heels. The only hurt was a broken or cracked collar bone.

When somewhat older, and quite old enough to know better, I cruelly tried to make old "Jule" the faithful old black, blind mare stumble and throw herself by riding her very fast across a little ravine that rose suddenly on the other side. She did throw herself but stumbled enough to throw the rider far over her head and beyond to the ground, striking my knee on a rock so crippling me that I could scarcely walk home. it might have been worse with me but I dare say I never attempted it again. "Served me right" I hear someone say.

Poor old "Jule." If ever a faithful animal lived it was Jule. She never flinched to do her duty and her side of the doubletree was always a little ahead of her pulling pardner. Indeed we always thought that her faithfullness (sic) in pulling caused her to lose her sight by becoming overheated.

Father never traded horses but kept a good animal as along as they lived. old "Jule" and "Nell" were the family horses as long as I can remember horses. The former died _____. The latter had heaves or consumption and after living a long useful life suddenly began to fall backwards and died as father was taking her thru the woods to pasture.

Since I am writing of the old family horse I mention that before father sold the "new house" they would cross the hill between our home and the above lot and for some reason pastured old Jule over there. Water was plentiful in the pasture and it would sometimes be several days before there was any occasion for a visit over the hill. I can well remember their scare and excitement when coming into the hollow and looking over the field (and) they could not see old "Jule." Upon closely hunting they found her down in a ditch in such a way as to be unable to get up without further aid. Without aid she should have perished. How father and mother were excited till they saw there was still life in old Jule. They hustled around and soon had her on her feet again not much the worse for her critical position.

Old Jule was responsible for another accident that caused me a bitter disappointment. Being a farmer, father always had some colts on our little farm. One or two of them were running loose and playing in the field or meadow below the barn. Another happened to get loose in the horse stable and had crept in the stall with old Jule. As she was blind and could not tell what the colt's business might be in her stall, she used her only means of defense and began kicking in order to chase the intruder out.

I was about to enter the stable and had one foot on the sill but tarried to watch the colts frolic in the meadow. Scarely (sic) had a (sic) placed my right foot on the cill (sic) when the left heel of old Jule struck the door jamb, slipped off and struck my right knee on the joint. Had the full force of the stroke hit my knee it might have been worse. As it was I was unable to walk to the house unaided. This kick altho you could scarcely see that the cuticle had been bruised or marked kept me in bed for about six weeks. My knee became very sore and could not be used.

As to the disappointment mentioned above I will say that I had learned to play the accordion pretty well, and I had an invitation to go on top the Allegheny Mountain to a picnic to play the accordion as the people marched from the school house to the grounds. So thoroughly was I making calculations for this great march that I could mentally see myself standing or walking at the head of the procession leading the mighty host. Then I thought I was to be the great attraction of the day as the musical prodegy (sic) of an accordion player.

Just that morning I had remarked to brother J.B. that I was going to be very careful that I did not meet with an accident so as not to be prevented taking the trip. But ere the day passed I had been visited with such an accident that kept me in the house for weeks. It was a sore disappointment. I was a mere boy and many times since have I wondered that the affair was planned. It certainly would have been a very strange procedure.

I recall too that I did not get to attend the "last" or closing day of our school and this was to have been a day for showing of my skill on the accordion.

One time I was throwing hay into the barn floor and after the amount was down I slid on the hay to the floor. I must have thrown a hay rake down before me for in sliding down it caught me in a fold of my "currying coat" as the coat was called, and ramming a hole through it, the sharp end of the handle caught me under the chin which was jerked up preventing the handle from passing up through my chin and impaling me on its end. This accident some how always impressed my mind with what it's fearful consiquences (sic) might have been.

I recall that on one occasion I was driving the cows to pasture that a rather cross bull resented the driving idea and tried to drive me. At least he turned on me and horned me striking me in the groin marking or leaving the mark of his horn on the side of my _____. it was a narrow escape that he did not kill me, but his horn slipped and I did not realize the seriousness of my position till I saw how scared father was when I told him of what had occurred.

At another time we were ready to begin thrashing with the thrashing machine. I was going to put the straw away in the shed when through carelessness I stepped between some polls (sic) that slipped apart and the result was that I fell some 11 feet to the ground. I was not much hurt but considerably jarred up. I think it gave me an opportunity to go to the house and rest for a few hours to see if I were hurt or not.

My first diary says that Sat. evening Nov 2, 1872 I cut my foot. I was displeased that I had to go out in the cold and cut some kindlging for Sunday's fires and said I was going to cut my foot, not meaning it however. Sure enough the axe suddenly slipped through a stick was splitting and cut a gash in my left

Undated tintype photograph of the Holsinger family at their cabin near Pleasantville, Pa. From left: William H. Holsinger (1871-1932), Joseph B. Holsinger (1858-1934), George B. Holsinger (1857-1908), Joseph H. Holsinger (1831-1895) and Rebecca Blackburn Holsinger (1831-1882).

Undated tintype photograph of George B. Holsinger
at the organ. This could be the organ that his father purchased
on Sept. 24, 1874 (see page 53).

Sallie Kagey Holsinger (1854-1930) in an undated photograph.

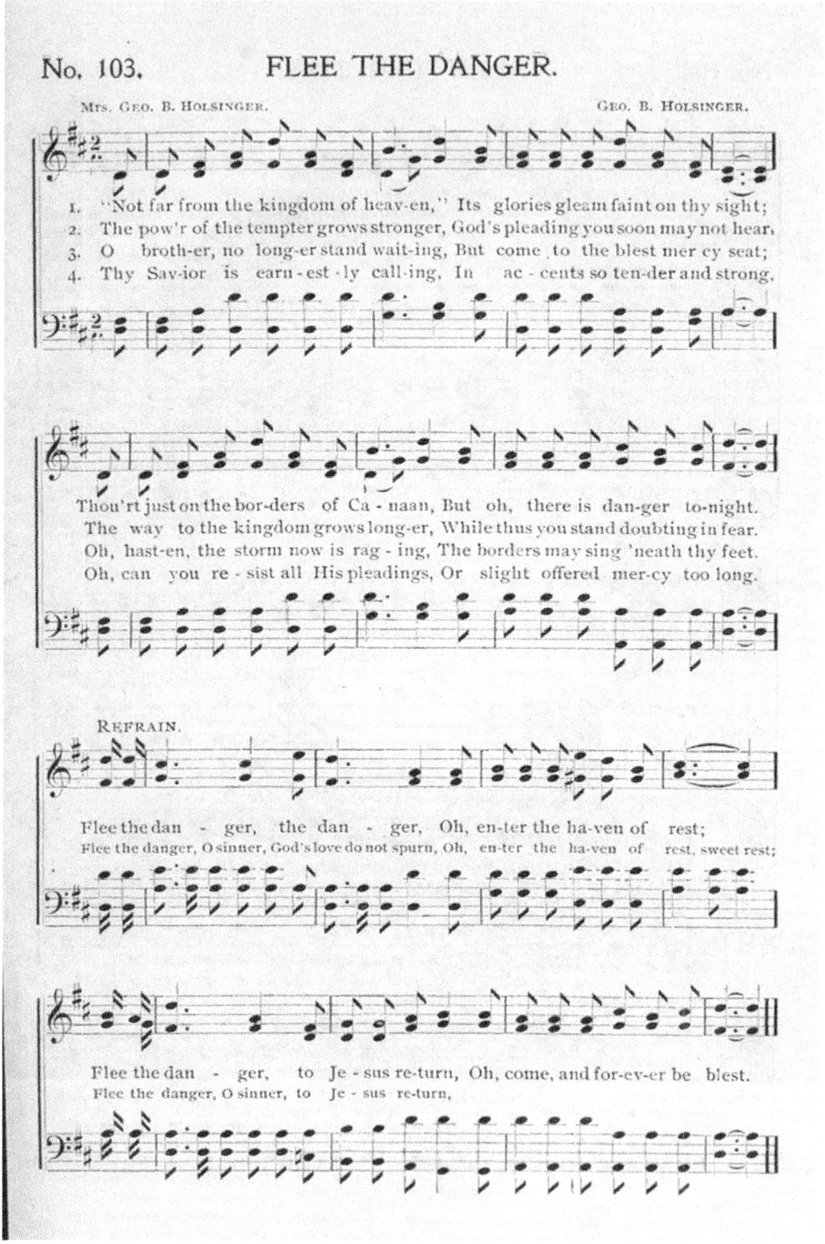

Flee the Danger, with lyrics by Sallie Kagey Holsinger and music by George B. Holsinger.

foot an inch or more long and pretty deep. This laid me up for repairs for several days and taught me a lesson not to threaten to do a thing I really did not mean to do.

The cut gave me considerable inconvenience for a couple years.

In 1873 when I was in my 16th year and being a lusty fellow of 163 pounds I was able to do some pretty heavy work. So in harvest time I could handle a grain cradle with considerable skill. Still I remember that there was something particularly exhausting in it to me always, and really it is hard work to handle a cradle in heavy grain. I had been working pretty hard and feeling like resting. I sat the cradle down with the sharp edge of the seithe (sic). I was awkward enough to sit right down so close to it that my right knee came in contact with the edge of the scithe (sic) making a wound that might have by a little more carelessness been very severe.

It may have been during this very harvest and as I now remember it, in this same field, that I learned a valuable lesson and one that father and Bro. J.B. must have had not a little laugh at my expense. I think that possibly there are few families where there are boys but that the boys sometimes have little disputes or in other words not to seem too pious, I should say quarrels. It need not be understood that were were an exception for we had our spats like many other boys. However as I remember it now this was about our last one and this was not serious. We were old enough to know better but we seemed not to realize this and so I proceed to relate an incident that I shall never forget. Father, brother J.B. and I were on the way to the harvest field but some dispute arose between J.B. and I and we kept up our jouring (sic) for three quarters of a mile when father, thinking I was the aggressor, which may have been the case, got tired of it and, siezing (sic) a dead lorrel (sic) brush gave me chase, slashing the brush around my lets pretty sharply before I made my escape. In all probability he could run as fast as I could and I may have gotten the full benefit of his just wrath. This only angered me but ended the quarrel. Being near the wheat field I had only time to resolve to get revenge out of father and bro J.B. by making them rake and bind sheaves as they had never done before or they would at least be left behind.

I felt strong as a giant! I whetted the scythe, walked to the foot of the hill—we walked down and cradled straight to the top that being the custom—I struck out, slash, slash, the grain fell into the cradle in great sheaves as it were. I got to the top quicker than ever before, but saying nothing to no one and not waiting for the sweating, (illeg.) binders to finish the swathe I had laid down for them, I again fairly ran down the hill and without whetting the scythe again cut to the top of the hill. This was repeated a number of times. I whet hurriedly and away I went. I suppose father thought the harvest of wheat would soon be over for never had he seen me cradle so swiftly and diligently. And I thought I was getting sweet revenge. But I was overestimating the strength of the giant. I do not now remember how often I climbed and descended the hill at the pace set at first but I do remember that I was very glad to hand the cradle to some of the other of the trip and my ugliness had been severed as had so many stalks of wheat. This was the last time that father ever needed to use the lorrel (sic) brush, or perhaps I should say ever did use it for the necessity or deserving to be used may be another question. in this case I had the pleasure (?) of "cutting off my nose to spite my face."

Some time after this but the exact date I do not know I met with one of the most severe accidents that ever befell me and came near ending my career.

Father had gone to "Chestnut ridge" for a load of lime. it was just fresh from the kiln. He got home with it towards evening and it must be unloaded that evening. I was the one to do the work, and possibly bro J.B. was to assist. At least he was present. I had thrown off a space about a yard square or more, making a pretty good size pile. On this father was carrying water from the horse trough some five rods away to slacken it. As boys we had often heard that slacking lime would get so hot as to set wood on fire. So we tried to set straw afire but did not succeed.

I then went to work shoveling off the lime. Boylike I would stick the shovel in the lime and rest on one foot. Then I would sway or swing my weight onto the other foot, and by this action I was not sure of the place I put my foot. I missed my step and fell into the pile of hot slacking lime. Having on short tow pants of my mother's weaving and making, and being barefooted, my left foot was terribly burned or scalded nearly to my knee. On the top of my foot it was doubly scalded, the skin turning off into the muscle, in size at nearly the top of my foot. This was caused by the second scald before I could get out which I did with my quickest effort, seeming to realize the peril I was in. I was scalded in other parts of my body but not severely in comparison to that of my left foot.

Brother J.B. set up a great cry and fuss that I was killed but I soon rose from the great cloud of dusty hot lime and steam and began walking to the house, remarking that I was not hurt, but looking to my feet I saw their scalded condition but walked to the house before any pain came to my foot. Finally it came in all the severity of such a terrible burn. My frame shook and I begged to be allowed to go to the water trough below the spring house and get relief. But they said I would take cold. Father poured lindseed (sic) oil on the burn which helped, and which was the proper thing to do. Finally I became too sick to sit any longer on the big chest on the porch and was taken to bed. This happening late in the evening they did not send for a doctor till next day. During the night I should judge that from the shock to my system I got a little rest but they had allowed themselves to be persuaded by me to fan the burns with a fan. This eased the pain but caused the sores to take cold and the consequence was a very severe sore. Dr. Headrick came next day and bound my foot in cotton and used lime water to heal the sores.

This was certainly my only salvation for its sediment helped in forming new cuticle. Six long weeks were again spent in bed and when I rose the new cuticle formed was too thin to hold back the blood which oozed out as through a cloth.

The wonder is that I ever survived the accident. My life was certainly despaired of by more than one person who came into my room. Cotton rags were kept between my toes else they had grown together. But I need not record all that was done and experienced. God seems to have had some reason to spare my life for many a person would have died with less severe burns. With God's care and my great vitality of body I was spaired (sic), tho even more severe accidents came afterwards.

An unlucky fox squirrel: Some time during the day, and before the accident occurred, I went up the lane and saw an unusual, large red squirrel. We tried to catch it but could not, so I shot it. Being the only game I ever shot and the last time I ever fired a gun. The squirrel was very large and no one had ever seen one of the kind. The gravy and meat were very good to me after I got an appetite. The fact of such a squirrel coming down our lane caused us always to think of the squirrel as an "unlucky squirrel." Superstition, of course.

Another accident befell me May 31st 1876. This was a cause of another disappointment perhaps less acute than the one caused by the kick from old Jule. Ever after this I learned never to make many calculations for it seemed whenever I did they were generally ended in disappointments.

On the above date brother J.B. and I went to Uncle Jos. Rouzer's to help him peal (sic) bark. He had just started a new home in the Rocklick hollow and being lone handed he had gotten consent of father that J.B. and I could help him several days on the clearing.

J.B. and I were there early if not arriving the night before, and we had cut down and pealed (sic) the bark from several large trees befor (sic) 10 o'clock. Aunt was needing some wood and so she suggested that we cut some wood till dinner and she would have early dinner and could go further from the house after dinner.

Uncle and I began cutting on the same log "old man fashion" as it is called. His axe struck a chip and glancing to the right struck my right knee just below the kneecap and a little to the left of same. The shock caused me to fall but uncle caught me. Bro. J.B. was soon carrying me into the house.

The wound bled but little. Our bark pealing (sic) was ended. It was thought best to take me home before the wound got too sore, so in the evening I was loaded up into a spring wagon and taken home.

It was a slow, tedious, painful trip, so we had every rough road to travel and had to cross a very high rough hill. Again my vitality through Providential guidance asserted itself and by June 25th or less than four weeks from the time the accident occurred, I had gone to a neighbor's house on a visit. On June 30th I slipped on the dewy grass, got a fall that caused the wound to bleed.

Had a dr. been on hand to have sewed up the wound it would soon have healed but he was not called for a day or two and it was then too late. Dr. Headrick did try to pull it together by some adhesive plaster but with poor success.

It was found that I would likely severely hurt myself and permanently become a cripple if I should remain on the farm for my knee was very sore, so on July 18th I was taken to Morrison's Cove to New Enterprise where I began my first experience of attending a Literary Normal outside the common schools. More of this later.

I must not forget to speak of the disappointment above mentioned, which was simply that by the accident I was prevented taking a visit to an old acquaintance who had two daughters for whom the father had just bought an organ. I was to have gone some evening during the week to try the organ. I had been playing nearly two years and was expecting to spend a pleasant evening. (My wound did not cripple me much as might have been expected).

In 1879 while on a trip to Mt. Morris, and while making hay for a man, I tried to do as I had seen others do, stick my hay fork into the ground. But striking wide of the earth I stuck one prong of the fork into the top of my foot so deeply that it left a lump on the bottom of my foot, showing that the fork prong must have gone nearly thru the bony part of my foot.

I did not get as sore as I expected, but I can remember how my thoughts went to mother and home, when during the long hours of the night my foot pained me and thoughts of blood poison and amputation flitted thru my mind.

I can recall a falling from old Nell's back one time returning from a trip to Pleasantville. I was in a hurry to get home and made her trot and she being a hard trotter and I a poor rider the saddle girth broke and I fell heavily to the ground getting a pretty good shaking up. I can recall at least two other similar falls with less of a shaking up.

My hands are full of cuts and bruises received from careless handling of tools.

In the month of May 1877 I was twenty-one years old and so Father proposed that if I would work for him at the farm till after harvest he would pay my tuition in another summer Normal school to be held in Pleasantville, Pa. by Prof. J.C. Polls.

This programme was carried out and another ten weeks Normal was attended. In the fall I was an applicant for a school in the home township, but did not get the school I wanted, but the directors saw fit to give me the most distant and poorest school in the district. But I accepted it and resolved to do my best. My first effort at teaching was, I may say, a real success. My pupils liked me, they learned fast, and what pleased the patrons very much was the fact that I taught the boys and girls some beautiful songs to sing. Even an old Quaker, with a son's family were most wonderfully pleased in everything I did and the music especially. Quakers as a rule do not approve or or did not then approve of singing in their public worship. On one occasion when I had gone to visit the daughter of a Quaker, she having attended one of my singing classes, I thoughtfully began to sing some of the tunes we knew. The room door opened quietly, the father stepped towards an out side door, no doubt with the intention of opening the door and showing me the way out.

But if these were his thoughts, which the young lady afterwards said she believed were his intentions, he evidently changed them, for he turned facing me remaking "we can do without any of that in this house, we don't want any more of that in here," walked into another room. I tried to apologize but the old gentleman was not in much of a humor to care for the apology.

The young lady was on the verge of stopping my vocalization several times but hesitated to do so through respect, but I certainly would have preferred her to do so, than to have received the stinging reproof of her father. The old gentleman would often attend our preaching services and show how I had some regrets that I must sing right in his face knowing that he did not enjoy it. I am sure I did not enjoy what might seem otherwise a means of getting even with him for his rather unkind reproof.

But to go back to my story of this my first effort at school teaching. My pupils attended regularly and when the last day of school came I had the unusual experience of having every pupil present and nearly all if not every father and mother of the pupils including the old Quaker and his wife. The pupils said speeches and I had rehearsed the day before and perhaps several times we sang the songs we had learned and that "last" day was the greatest day of the school at the "Upper Barefoot Hollow" as the school was named. There were several families by the name of Barefoot living in the township and so there was a "Lower Barefoot School" too.

As an evidence of my success with this school I may say that the citizens offered to board me free if I would return and teach the school the next winter.

But by this time I had made such a reputation as a teacher that the directors gave me the school I desired the next winter. This time it was the "Moore" school or the very school that twelve (?) years before I had entered into a wild and untried new experience of the first day at school. My success was again quite as great as it was in my first school. At the close however I was not a little disappointed that one family did not attend the "last" day for I was secretly hoping that as at the close of my first school every family, parents and children would attend. I have never learned why the family did not attend but think there was just a little intention on their part to disappoint me , if they did not know, even that I was hoping to have all present.

However they missed a very enjoyable event to all present for I had prepared a number of excellent songs and to make the music more interesting father had permitted me to take the organ to the school house and I played the pieces that were sung. He hauled it for me and he and mother were also present. "Speeches" were made by all the pupils and really the day was one that I dare say none of the pupils ever forgot. I knew I was very happy at the way all the parents and pupils of all the families of the district, with the exception, turned out. The most aged persons in the district men and women were there and the house was packed to its utmost capacity.

As said before this was the place where I first attended school and had seen different exercises at the close than what I had when I taught. I had seen the young men and women dance and play in a way that I had been taught to think of as very wicked. There was a confusion and a carousing that was quite different to that of our closing day.

I had been asked to teach this same school the next year but the directors say fit to give me what they thought was the best school in the township. For some reason another teacher also wanted this same school and the disappointment to me of not getting the school I wanted caused me to exchange the school allotted to me for the one at the "Upper Barefoot" hollow, which he was to have taught. However the citizens while they were glad to have me return were not quite so enthusiastic and in a sense had forgotten the offer to board me free if I would return the next year. Still all but one or two did their share toward boarding me free. Some of the pupils had gone and I found it hard to get up the same interest among the pupils that they manifested the first year. In fact I must record the fact that the "last" day was quite a tame affair to that of two years before. There might be reasons given for all this but perhaps it is unnecessary to do so.

I had a little experience towards the close of this school that

I cannot help recording. it was the pulling of the first tooth ever I had extracted. The tooth had been aching for some days and was very annoying, so I resolved to stand it no longer. There was a blacksmith living just a little way from the school house so at recess I went to see if he would do the job. I think he was quite willing to do so and possibly he thought he would get a little sweet revenge out of me, for he was displeased somewhat at me for whipping or punishing his youngest son. He need it or I should not have administered the chastisement. He sent one of the children to the house for his instrument a crude "cant hook" affair. I handed him my handkerchief and he wrapped the same around his terrible instrument so as not to lascerate (sic) my mouth too much. I had never had a tooth pulled and had no idea whatever that it would hurt to have a tooth pulled any more than it would hurt the earth to pull a stump from the ground,.

Imagine my astonishment when the blacksmith gave a wrench at the tooth which flew we scarcely knew where, for we had to hunt sometime before it was found, at a feeling that my head must surely have busted asunder. I am sure nothing had ever happened to hurt me so suddenly and with such excrusiating (sic) pain. Indeed I could not imagine what had happened. I think little was the teaching I did after recess and that night I had all kinds of dreams for I thought or dreamed some one was piling heavy logs on me.

My jaw was sore for months, for I think the jaw was cracked, the tooth being a _______ in the upper jaw. To this day I sometimes am troubled with neuralgia where jaw was injured.

By some means I was persuaded the next year to accept a school at the little village of St. Clairsville. I think at came through the persuasion of a friend one of the directors. The school liked to prove my Waterloo. I was brought up in the country and was not used to the way of town boys and girls. There is a ferlessness (sic) a badness in many cases that country boys did not manifest, at least those I knew did not. The school opened, in the little old log house, used that winter for the last time. The house was really too small for the 64 or more pupils who attended regularly. The first week, possibly the second came and went, but some how I had not give into that school with the right kind of resolution to succeed.

I seemed to be unusually timid, scared and bashful. I was afraid of the ways of the townsfolk, and always was unusually bashful. I should have mentioned that in the 64 pupils were some 18 or 20 young ladies and nearly as many young men. Whether they noticed my embarrassment I do not know. However, they were easily interested in their studies and I succeeded with them admirably after I once resolved to finish the school. But there is a little something connected with the forming of this resolution that has always had a little something to do with my success in life if success be recorded.

After the first or second week which were terrible lonesome long weeks, without saying a word of my intentions I took my clock, my books, and an all my belongings to the office of some one, so that I could get them without having to go to that school house for them. I had fully resolved never to enter that school house again as a teacher if I could help it. I started homeward. A heavy snow had fallen on which was a hard crust of ice so that it was just as easy to walk on the ice across the fields and thus make "shortcuts" as to follow the road. I stopped that night at my uncle Thomas S. Holsinger's some eight or more miles from St. Clairsville. I told uncle of my discouragement and how that I had resolved never to return to finish the school. Uncle in his characteristic way said, "George, be a man or a mouse."

I could hardly understand his expression but he soon explained it by saying that I had been successful, and quite successful in previous schools, and if I would fail in this one I would be so crushed as never to succeed thereafter. My reputation would be ruined. On the other hand if I succeed now that I had the greatest opportunity ever offered to me, my reputation was made and I would after that be able to get any school I might desire. His cute saying, with its explanation was a great source of encouragement. I went home Saturday morning and after talking over matters with Father and Mother and receiving their encouragement and spending a rather blue Sunday, however, I resolved to go back and "try again."

I think it took some weeks however to get the resolution fully grounded and rooted to finish the school. As a whole the school so far as the part I had to perform, that is that of teaching could certainly have been considered a success, for I could offer some very strong evidence from what young men, and I now recall the evidence of a young man since a very successful preacher who confessed that what I did for him that winter was the inspiration to him that gave him a start, and made his life a success.

But with all this I was too glad for the closing day and did not attempt any special programme. That closing day was the same as any other day.

I happened to have a boy or two—brothers they were, that were unusually hard to manage. Trouble was coming and so crossing the field I usually crossed in going to the house, I cut some hickory switches or as I called them whips. One day the older one was so unruly that I concluded that I must administer some hickory oil which I proceeded to do, I thought I have a pretty heavy dose but imagine my surprise on turning suddenly around, to find the young man standing up straight pointing his finger at me. I proceed (sic) to administer a second application of the hickory oil and doubled the dose. This time the young man did not do any pointing. I had the control the my temper so well that I never seemed the least bit angry when administring (sic) discipline. This time as usual I was perfectly calm and it had a good effect on the school. That boy never gave me any more trouble but his little brother thought to take revenge by being naughty, and helped I think by his mother tried my patience sorely. Once when about to punish him he slipped out of the doors before I could get to him. His mother soon appeared on the field with a story of a weak heart, and that he did not dare to be punished, but I replied that his heart was not too weak to be bad and that he could not return without receiving just punishment. As I was firm in my decision I think he did not return.

The school was well advanced and it was a good schooling

for me, and I had to study hard to keep ahead of the pupils. I remember that on several occasions I had to call on a well educated minister for some needed help in grammar.

I happen to have a vivid memory of a very amusing incident that happened to me on one occasion when I was crossing the aforesaid field. I was going across the field bound towards my boarding house, with my arms piled high with books, on top of which I had my clock and in the other arm my dinner pail. I was walking slowly and thoughtfully when suddenly and without any premonition my feet suddenly flew from under me and I seemed to sail away into space. However there seemed to be received from the direction I was walking a very helping push, and if I had been walking on a railroad I might have concluded I had been struck by a locomotive silently creeping upon me. Books, clock, etc. flew in all directions and upon recovering my equilibrium and turning to see the cause of my sudden flight skyward I found an old ram had slipped up behind me, delivered a heavy blow and Merimac-like had backed off to try a second administration. I do not know why I was carrying so many books but it may be it was on one of the occasions when I had notions not to return to the duties of that school. The old ram may have divined my intentions and desired further (?) to administer a rebuke for my sheepishness or to have helped me somewhat in my resolutions.

During the summer of 1879 my uncle C.S. Holsinger was asked to make a visit to Illinois to visit the contrat (sic) Illinois Mission Field with a view to taking up the work in that section of the state as a mission worker. I being then 22 years of age, and not being specially needed on the farm at home, and uncle desiring the company I would be to him I concluded to accompany him to the great Western country. My uncle Jno. S. Holsinger thought it would be a good thing for me provided it did not make me a rover and so unsettled as ever after to desire to be traveling, as is so often the case. Altho it has been my lot to do much traveling, I do not think the trip had the least to do with it.

I bade a fond mother and father "good bye" and we started to

Johnstown, Pa. where we were to take the train, May 25, at 9-30 a.m. so says my diary. The account of my trip is rather full and complete, but it is interesting possibly to me alone.

It was my first car ride and of course there were many things to interest me and attract my attention. One I note was that "a little man 28 inches high, dressed in broadcloth high plug hat, broad shoulders, auburn chin whiskers etc. passed thru our car. It was none other than "Tom Thumb" and across the aisle and a few seats back of us sat his wife. The first card to mother told her about him and what sights I had seen.

The number of people traveling, the sights of a city smoky and dirty, the first boat ever seen by me, sailing on the waters of the great Ohio river, the dizzy onward movement of the train and all receive a passing description as we move along. I think I must have gotten little sleep that night from the notes I made concerning the places we passed through.

Arriving at Mt. Morris at 3 p.m. we were taken to a love feast at Silver Creek church. Stopped all night at Bro Benj. Swimngley's. We visited some of our relatives of the Holsinger name, Mt. Morris College having just been purchased, and the school was to have started that fall.

We attended communions at Silver Creek, Rock River, Mulledgeville, Pine Creek.

We finally got to our destination in (illeg.), [Mc]Henry Co., where we remained about two weeks.

During our trip here a rather amusing incident happened thru which Uncle C.S. and D.L. Price had considerable fun at my expense. The subject of using tobacco came as we were going-driving to church one evening. When some one of the party to break my silence asked me if I used it. I remarked I had never done so when an old lady, whose daughter Mary was also in the crowd, clapped her hands together and shouted, "Good for that, Mary always said she never would have a man who used tobacco." I had not been paying any attention to Mary and have strong resons (sic) not to do so, even though the way had been so nicely prepared to do so.

Returning to Mt. Morris Uncle soon went home and left me

in the lonely west. I made my home at Eld. D.E. Price's working on his farm and for my cousin and others. Work was done in such a different way than what it was in Pa. But I rather enjoyed it. But this was my first trip away from home and I should record that I experienced my first "home sickness." Altho (sic) I had a good chance to stay I prefferred (sic) to go home and so on _____ I started for home. Poor Mother when she heard that Uncle C.S. would arrive on a certain day walked three miles to meet us but I was not there and then I happen to know she went home with a heavy heart. Her longing for her boy was expressed in her letters and they had quite a drawing to pull me home when she used such expressions as "O, George come home." I once heard her say that she did not like our old black buggy for it took me away so often.

I got back too late for a school in our home township so as St. Clairsville school had not secured a teacher I was induced to accept that school the following winter.

I now come to the summer of 1881 during which an incident happened that I have always considered as the turning incident in my life. But for this there is no telling what my life would have been. Surely I would have worked along a different line. Possibly I would have been a Pennsylvania farmer.

It is possible that I attended an invalid at Dr. Watter's Mountain Park Home, near Wearnersville, Pa., and got started to Normal School at Pleasantville under Prof. C.J. Polls four weeks after the school started. This was somewhat against my most successful pursuit of my studies. I had been studying instrumental music without any teacher and had attained considerable skill, or at least skill enough to be able to play at sight almost any athem, Sunday School or church tune.* By this time I had even conducted some singing classes.

Prof. Polls always had a Literary Society connected with his

---

* Here Holsinger wrote lightly, in pencil, "often ballads."

Normals that met every Saturday evening and this may have had something to do with my having my organ in the hall. But we always had music in Prof. Poll's Normal and this fact may also have had something to do with my having my organ there.

I remember however that the Literary society spoken of above had beens tarted some years before and that I had taken an active part in it appearing the first session as I now remember it in a debate in which our side won. Note this fact.

Sometime during this session at the Normal I was placed on the program, to debate this same question and my opponent was a young man who had failed to appear on several occasions against me and so for some reason or other I concluded he would not do so again, and that there would be no debate and therefore there was no reason for me to make any special preparation for this occasion.

This young man, a lifelong school mate but a little younger than I, had for some years been rivals in the matter of getting the higher grade Teacher's certificate, but I kept a little ahead of him, so when he appeared unexpectedly on the field of battle, with a speech well prepared and began to deliver it in his characteristic way, and knowing that I was not prepared as I should be, I confess that I was so completely humiliated that I was unable even to say what I otherwise might have said. I refused to say anything and some one else had to take my place.

To make matters worse I had another duty to perform in the preparation and reading of the literary weekly paper. in this I should say I was also direlect (sic) and receiving no contributions from those appointed to prepare articles, not a line of the paper was prepared and the embarrassment of a few minutes before was repeated. I was so much humiliated by the first failure that I never thought about the paper till I was called on to read it. All I could do was to explain that no contributions had been received and therefore I had failed to prepare any paper. The old saying that "calamities never come singly" seemed to assert its truthfulness towards me.

The house was unusually well filled at this meeting and some company was there from the west, who were visiting one of the principal families of Pleasantville.

Someone must have had pity on me, and felt that I was receiving an unjust amount of humiliation, and also that I should have a chance to redeem myself by doing something I could do, so it was proposed that inasmuch as I had failed in the other matters that I should either play an instrumental solo or sing a song.

Prof. Polls came to me quickly and whispered in my ear that I should play the ballad "Will You Have Me When I'm Old," which I played and sang as I think I never had before and possibly have never done since. I seemed to receive an inspiration that helped me wonderfully and strengthed (sic) by the confidence that I was going to do something I <u>knew</u> I <u>could</u> <u>do</u>, made this my third task an easy one. The song was wonderfully well received and such a burst of applause as I received more than recompenced (sic) for the humiliation just passed through.

I think it was before I left the Hall that I had resolved (then and there) to make music my life work. At least I know that I remember well that on the way to my brother Jos. B's home, and with whom I was boarding, that I told some one of my intentions. I immediately wrote to the manager of a Musical Normal school to be held at Dayton, Va., and in a very few weeks I was on my way to the above place.

More about this Normal and what it resulted in will be told elsewhere. But that the above incident was a turning point in my life I do not hesitate to confess.

My father was as great lover of music, always "raised the tune" at every meeting he attended, for he had I think quite the best voice I ever heard for a leader of singing. It was not loud but he (it) seemed to possess a quality that could make itself heard in any audience, no difference low large. Besides having such a splendid voice he was also gifted in the way of knowing very many times so that he never was at a loss to select a tune best adapted to the hymn to be sung. This gift was perhaps the result of a special preparation or effort on his part to know all the best hymn tunes, and to have them in readiness on all occasions where he might be called on to sing. I do not remember ever hearing him make a mistake in starting a tune not having the right meter. in his day a tune book with all the tunes already adapted to the hymn was not known or at least not till H.R. Holsinger compiled his "Hymn and Tune Book." Even this was a crude affair and was not strictly adhered to by father. He chose rather to select his own tunes to any hymn selected by the minister. Father never used a "tuning fork" to get the right pitch but seemed to know it if he had pitched the tune to (sic) high or low and had the skill to slide his voice up or down as the case might demand. This after all is as it should be for he was simply using his musical faculty in the same way that the generations past had judged of the hour of the day or night by the using of the faculty given by an all wise Creator for the purpose of reckoning time. Today instead of using these God given powers men use the tuning fork or pipe and the watch, invented like so many other mechanical devices to do our thinking or as in many cases do the work of human hands.

Boy-like I thought father was the greatest singer living, and how I would listen to his voice as a bass singer on occasions when he did not happen to start the tune. I wondered if there were any other persons who could sing so well as could father and I was very proud of him to sit by his side on these occasions.

While father read the Bible as noted elsewhere still I think the greater part of his spare moments were passed in learning some new tune and he had to work weeks, possibly months, to learn them well and to be able to sing them when they were wanted.

Once learned he would seek the first suitable opportunity to sing them at public worship. He did not do this to show that he had learned a new tune but that others might have a chance to learn it also. That was the only way any new tune could be learned in those days.

I recall how that mother often tired of hearing father learn new tunes, for it took him so long and he would repeat time and again some part of the tune that he was learning. I recall at this writing one especially that required a special effort on his part. It was to tune written by A.D. Fillmore to the words "We speak of the realms of the blest." I have not the least doubts but that father and mother are now both enjoying "The Realms of the Blest" for both have been dead some years.

Mother liked music but she had no special gift or interest in it altho she could sing and I suppose it was through these facts that she allowed herself to tire of father's continual singing.

Whether it was thru hearing father sing so much or that I had musical talent transmitted to me I am unable to say, perhaps a little of both, gave me the interest in music I seemed always to have.

When but a very small boy and certainly before I can remember of it, I was very proud to have the privelege (sic) of taking part in the marching that was often done while singing. This may have most likely been through the faculty of imitation largely developed in children.

Be this as it may I have heard them say that singing always had a very great soothing effect on me, whether more than on most children I cannot say.

As I grew larger I know that somehow I thought I was somewhat of a singer and once when it was said that some one in one of the classes was to start the tune I thought that must mean me. But I do not remember that I did so, but <u>do</u> remember that I always tried to sing when any singing was being done.

Father always had an old French accordion on which he could play the tunes he knew and on this I learned also to play. He finally sold this and bought what was called a German accordion. Then a larger one was purchased. Indeed it was the largest accordion I had ever seen and I do not remember having seen a larger one since. This was sold and I bought a smaller one. In fact I believe father also bought one so that we had two of them in our family and a neighbor boy also bought one, all in the same key. On these we learned to play in concert quite skillfully. I have already mentioned in the "Chapter of Accidents" some of the rather amusing intended performances. Others I dare say might be related.

One day father came home from Pleasantville with the word that the store keeper had bought an organ. I do not know that I knew what an organ was or that I had ever heard of one, but he told me how it was made and how it was played, and that he had heard it played.

I knew that I could hardly wait to go to town to see and hear the great new musical instrument.

The next time J.B. and I went to Pleasantville we asked the store keeper about his organ and if he would not play it for us. This he kindly did and I am sure I cannot describe my feelings. That so much music could be put into so small a box was to me a most wonderful revelation. It seemed to be I never before had heard anything so musical. It was a new world I seemed to be in. I think I felt that I never would be quite satisfied till I had an organ and could learn to play on it.

Sometime after this there was to be some kind of a political meeting in Pleasantville, and we heard that there was to be a "brass band" there. I think that Brother J.B. and I went alone to hear the "band." As to the speeches that were made I remember nothing, but I remember that some seats were erected for the liseners (sic) and especially for the band. Finally the meeting was over and we had to go home but I had been so thoroughly intoxicated with the music that I did not know where I was. Everything seemed strange and had turned round so that it was several minutes perhaps much longer till I realized where I was and everything began to right itself. My mind seemed to be so full of the "toot, toot" of that band music that I fell into the habit of making a kind of tooting noise or music all day long, that became so tiresome to father that he tried to buy me off not to sing or make that tooting noise. Persuasion seemed not to reach me.

Some time after this we heard that a man had just moved into the neighborhood where my Uncle T.S. Holsinger lived and that they had an organ for their son, and that he could play.

A visit was soon planned to my uncle's, but this time more especially to see the organ. The boy was somewhat vain of his accomplishments for he could actually play a very few exercises from the instruction book. He could play only the exercises from the instruction book. I was quite anxious to get my fingers on this organ, and as soon as he was done I asked permission to try to play. Imagine their astonishment, when they knew I had never tried to play, to see me strike out and play two parts, soprano and bass of the old tune "Sessions." I first fingered at the soprano and the bass alone and then put the two parts together. Incredible as this may seem to the reader, nevertheless I am telling real facts. My mechanical ideas, inheritted (sic) as it were from father and mother, gave me a mechanical idea of how that one played on the organ.

I think father got ashamed of my long effort feeling that I was staying at the organ too long, so he had to command that I go away. I obeyed but had the pleasure of playing again before I left. I made several other trips soon after this one and on one occasion heard a lady play that was really gifted and played skillfully. All this simply added combustible material to the already burning desire to own an organ.

On another occasion a Hall was to be dedicated in Pleasantville and altho father rather objected to our going undutiful as it may seem I could not resist the desire to go. A man by the name of D.N. Croft from Woodbury, Pa., had an organ at the dedication. After the dedication it was placed across the street in some house. I lingered around that house all afternoon. I was rewarded by hearing several persons play some fine music. One man especially played a piece that sounded to me as I thought like the whole brass band I had heard some time before.

Somehow I couldn't get away from that house any distance and when I saw a covered wagon drive up and the organ being loaded thereon I was not a little anxious to know where they man might be going so late in the evening. Someone not recognizing me said he was gong into the country to see a man by the name of Holsinger. I thought I knew all the Holsingers and thought he must be going to our place. Someone knew me and said, "Why here are Holsinger's boys and they will show you right where you want to go." I was greatly pleased to have the man go to our house with the fine big organ. We got home a little before dark. Father permitted the man to take the organ into the house, well knowing that it meant some good music. A great aunt had come to our house that evening, also one of the boys. Some neighbors gathered in and the man being a good singer as well as player gave great enjoyment to the company present by some of his comic songs.

Next morning Father consented to buy an organ. it was a happy day to me, the greatest joy of my life.

But the man was not so much in a hurry to bring the organ as I was that he should do so. Days and weeks seemed like months and years. Finally the organ arrived. The man did not bring the organ we had bargained for and so father was slow in paying $20 more for the organ than he was to pay. But after parlying (sic) and by my urgent persuasion he finally did by (sic) it.*

Thus began my practice in earnest. in less than a week I had learned to play more pieces than some others that had had an organ for months.

I should not forget to relate that after the organ had been bought, and during this terrible long month, I had marked on a single the keyboard of an organ and practiced on this imaginary keyboard, which no doubt was no little help to me in the light of the techniques so much used today by learners.

During this long month I no doubt worked very steadily on the farm but when a week had gone by I began to think that soon the organ ought to arrive. So every time I would get to the end of the furrow or in sight of the house I would cast a searching

---

* Penciled note in margin of the manuscript notes that the organ was purchased Sept. 24, 1874.

eye to see if that black covered spring wagon might not be seen driving up the lane or into the yard.

Frequent excuses were framed to get to the house and on one occasion I was plowing and a trip had to be made to the house, so I went but did not tie carefully the team and on my return I found our best horse "Tobe" had fallen, striking his shoulder on a sharp point of a rail fence. The wound was so sore that we came near losing him. My diary afterwards mentions the fact that we sold Tobe for $200. So much for my being rather too much interested in something else rather than the work in hand.

I may add that every spare moment and some that were not "spare moments" was spent at the organ. My progress was marked. I soon began to play three parts then came the fourth till at last I remember telling father once that I believed I would soon be able to play the four parts to every tune I might play. And it was, so for every key has more or less of a relationship to every other key and altho I did not know what to call the combinations of tines, I was really constructing as A.N. Johnson has so aptly put it, "a system of chords of my own." This is did and altho I did not know that the name chord should be given to these combinations, I was in reality forming a system of chords, and all that was needed was the teacher that could systematize "my own system of chords." In this connection I might mention that when I attended my first musical normal in 1881, that when my teacher Prof. H.J. Showalter had explained the tonic, dominance and sub-dominant chords, and had given us a melody to harmonize that before the class had copied the melody I had copied it and harmonized it. My teacher said, "Never mind, we will give you something to do by and buy," and he did.

I had had the organ not more than fifteen months till a friend who had just gotten an organ said, "I want you to give me some lessons." But I remarked that I had never taken any lessons and how could I give lessons. But she remarked that I could certainly tell her how I played and if she could play as well as I could she would be satisfied. I undertook the job and I can well remember the $5 bill she hand to me, being the first money earned in this way.

I cannot tell now whether my pupil so soon learned all I could teach her or whether she thought she could learn it herself or whether the winter weather prevented my giving the full number of lessons but perhaps the latter. At any rather I remember that I had the pleasure (?) of refunding some of the $5 paid in advance.

From what I have already said concerning my first studies in learning music, some might infer that I had really great talent in music. I know not what others may think but I put myself on record by saying that I do not think so. I believe I greatly loved the sweet sounds of music, but have often wondered if after all that great incentive to action that prompts the human family to do or not to do certain things was in my case at the very bottom of my industry in learning music. I refer to what the phrenologists call "Approvativeness"* or "Ambition."

All of my life I have found the promptings of this sentiment so strong that I have had to restrain its natural tendencies. Now it occurs to me that I had a strong desire to be able to play well, really well, so that it might be said of me that I could play anything or was a fine performer. I may be mistaken in this but when I reflect that I was not learning music with the idea of ever being a teacher of music.

I cannot but believe that I am not far wrong in my beliefs.

About the time that I became interested in music, my brother Jos. B. began his tinkering at watches. While I spent my spare moments at the organ he was fixing some old worn out watch. He inherited a wonderful amount of mechanical skill, and I do not believe there are many more talented mechanics than he.

The first watch he ever worked on came into his possession lacking a _______ wheel. He had no tools so he made some.

---

* Probably approbativeness, an excessive desire to become the subject of approval or praise.

To make a wheel he must have a very narrow file to make the cogs. He took one of father's three-cornered files and ground the edges to the right thinness on the grindstone. Then with an old German silver dime he made a wheel. He worked away at it till the watch kept good time. This was really a great feat for a boy who had never seen a watch repaired. Years afterwards this same watch came back to him with this same homemade wheel therein and he got permission to replace the old wheel with a new wheel. He discovered that his wheel lacked two cogs of having the right number. He still possesses the wheel.

When the organ was to be bought, J.B. and I almost quarreled about who would be able to learn, and learn the fastest. He believed he would and could learn to play the organ quite as well as I could. But his talent was in another direction, for I remember one morning of seeing him try to play. I could not understand what he was trying to do, for he did not succeed in making a tune go, and I really thought he was simply playing with the keys. But I afterwards discovered that he was really trying to demonstrate that he too could learn to play. I never saw him try again.

While I would use all the time I could spare at the organ and have nothing to show for the time, he would work at a couple of watches and immediately earn his 50¢ or $1. He has made watch repairing his life work and tho not realizing it then I was preparing myself for my life work.

I write the above to show how sometimes if left to follow our inclinations we may find our life work.

My first singing school teacher was Benjamin Holden Everett. He came to our home one extremely cold night Jan. 9th 1875. He came to our place knowing that father was the leading musical spirit of the neighborhood and of course desired his help to raise a vocal class in the neighborhood. Several schools were made up for him and he began his session Wed. Jan. 13th.

B.H. Everett was a one armed man and did all his blackboard work with his left hand. I have some very vivid recollections of his teaching and of this the first singing school I ever attended.

He was not greatly gifted in music and all his teaching was very mechanical, that is he had committed every word he would say to the class and followed his "system" to the letter. I can yet see him, draw his lines on the blackboard for he did not have them painted thereon, and it was surprising how he could draw such a straight line across the whole length of the board.

I met him only once after this class and that was in 1883 after I had located in Bridgewater, Va.

Before this singing by Mr. Everett I notice in my diary's (sic) that as early as _____ I lead in social singings but after the above mentioned class I did more of it. My first pay classes were taught in 1877 or 1878, altho I did not teach a great number of classes. But we had many meetings at the Dunning's Creek church.

I recall that at one time I had three classes at the same time and that some teacher in this case I ought hardly say he was a teacher, but a man visited one of our neighbors who thought, I imagine, that I was making myself just a little too prominent musically, so he thought he would have a singing appointed for this friend and if possible get a class for him. I had met all my classes just previous to this Sunday and told them there would be a singing on the above named Sunday by a Mr. ____ and invited all who could possibly do so to be present and to sing too if they came. The Sunday arrived, all my classes were were well represented, and the book I was using was in the hands of all my pupils. The man sang a while and I think he found the singing better than he expected for I remember he soon had some very pleasing things to say concearning (sic) how well the people sang. He said among other things that he did not care to interfere or to have a class, seeing the progress the class was making under the teacher who had taught them. In this he was very considerate and I think I was much flattered.

I happen to remember an expression of Bro. J.B. that voiced the feeling of the classes better than any thing I can write. He said, "I never sang so hard before," which showed that the people were just a little bit proud or satisfied, or jealous, of their teacher and wanted to show the intruder that they had been taught to sing.

I took my first instrumental or organ lessons the spring of 1882 at Max Beekhoeffer. While I was teaching my last school which was at Clayburg, Pa., or in that neighborhood, I did some teaching of the organ and had a chance to get a good class in instrumental music but thought that as I had never taken any lessons that possibly I ought to take some lessons, so as to be sure I was teaching as I should or in the best possible way.

I did not long continue, however, for my teacher simply had me play over pieces I had often played before and so there seemed little for him to do and he did but little. I may have needed the instruction but it seemed to me he was trying to hold me back so as to keep me as long as possible.

My next instrumental lessons were taken in 1881 (W. Funk) and also in 1882 (W.E. Burnett) and again in 1888 of P.S. Mergis. I also attended Theo Presser's great musical Normal in Phila., Pa., in August 1894 of Chas. W. Landon.

Along the line of vocal music my first teacher as said elsewhere was B.H. Everett and I notice my diary speaks of another teacher having taught in our home church but I have no memory whatever of him.

The next instructions received was, as said elsewhere, at Dayton, Virginia in the Normal of 1881. Here I received my first real teaching. We had studies in theory, harmony, composition, voice and experimental teaching. During this Normal I composed the first song I ever had published. It drew first prize and it was named "Sweetly We'll Sing," words by Aldine S. Kieffer.

The same Normal was attended the next year and again my song "Shall We All Meet At Home in the Morning, or, Gathered Home," drew first prize.

At this Normal I was organist. What occurred between those two Normals will be told elsewhere.

One event was the sorrow caused by the death of a dear mother which occurred in April 1882 and about which I write elsewhere. The other was a mere passing view of the lady who was to become my wife in three years. Of this I may write in another chapter. And still another event that has had no small _______ in my future corse (sic) and especially had to do with my connection with Bridgewater College was the meeting with Bro. D.C. Flory, then principal of Spring Creek Normal, afterwards B.C.

After this first Normal was over a man by the name of Wm. Steele engaged me to teach in a commercial cshool he was then planning. But as a few weeks were to elapse between the Normal and the opening of this school I was solicited or employed to teach instrumental music in the home of Jos. Click, living near Bridgewater, Va. I taught for ___ weeks and hearing nothing from the proposed business school and having no other engagement, I concluded to go my home to in Pa. About the time I was ready to start Bro. D.C. Flory happened to call at Bro. Click's. We were introduced and after he found out my business he said he was hunting someone to teach music in his school and would be pleased to have me do so. I was anxious for just such a chance for I could in this way pay my way thru school in the literary department.

Bro. Click made me a kind offer that would have greatly aided me in the proposed scheme, but it was not to begin that fall. As I had determined to go home I did not change my mind but went. I was too late again to get a school in our home township and no where else that I then knew of, so I had about decided to return to Va. One day I went to Pleasantville and after attending to the business in hand I was met by Bro. Henry Brumbaugh of near Claysburg, Pa. who was in the junt of a teacher to teach their public school. I was not at home when he got there but he told my father and mother what he was after, and my dear mother was particularly urgent in charging Bro. B. to persuade me to take the school, fearing as he told me that I would go back to Va., and so far from her and home should I not take the school. perhaps that very morning I had decided to return to Va. Bro. B. said if I would take the school he was sure that I could get as many instrumental pupils as I could teach besides some vocal classes. I agreed to take the school and it proved to be one of the best schools I ever taught and I have always been glad that I consented to teach it. During these 25 years that have elapsed since I taught that school I have received many sweet letters from pupils who attended that school in which they refer most kindly to the inspiration and encouragement received at that school.

As in all my other schools I taught the pupils to sing and this kept the school in a good humor and was a great source of enjoyment to all and I have had many precious assurances that my last session taught was fully appreciated. Twenty five years have since elapsed and now and then comes a letter laden with the precious words so much appreciated by every teacher.

It was during this winter that I had further correspondence with Bro. Flory relative to joining his faculty and before the school closed he sent me a catalogue for the Va. Normal for 1882-83.

It was also during the last week of this school that my mother took sick with pneumonia and after suffering only about a week she died April ___ 1881.

My school closed on Friday the ___ of ___ and not having ever been to visit the city of Altoona and having some kinsfolk living there I thought it would be restful and pleasant to visit the city on the Monday following.

On Sunday a friend took me to visit a friend of hers, and after a pleasant visit there seemed to come to me a desire to return earlier than necessary to my boarding place at Bro. Gochenour's, at ____. I went and hitched up and we started for Bro. G's. Before we had gone far, we were met by a cousin of mine, E.R.H. and a Bro. B. Greetings were exchanged and then came the sad intelligence that my dear mother was sick and that contrary to her wishes they had sent for me.

My friend said I will never be paler when I am dead than I was when I was informed of the illness of mother. This was Sunday evening and a Tuesday about three o'clock she died. This was the greatest sorrow of my life. No one can realize the sorrow of losing a dear mother till it comes to them.

When only ____ years old I lost a younger brother, David B., but my young heart was not touched as it was at mother's death.

Mother was sick only about a week. She took sick about Monday or Tuesday before my school closed on Friday. Mother knew I had worked hard all winter, and knew of my intended trip to Altoona and so she would not allow them to send for me fearing it would spoil my pleasure trip. Otherwise I would have been at her bedside as soon as it was realized that her sickness was serious. As it was I was with her only three or four days. She had wished for me several times for she said I was so strong that I would be able to turn her so easily.

How closely I staid (sic) at her bedside! And I would go to her bedside many times at night hoping that each time there might be some indications that she was better. How vividly it comes to me even now that on the last morning she lived I waked my brother Wm. H., saying to him that he should get up and see his mother, that this was certainly the very last morning he would ever see her living. Thus it proved for she died that afternoon.

I somehow always had an impression that my mother was certainly more concearned (sic), more solicitous about her boys than any other mother. I cannot see how a mother could be more so. But I must here relate an incident that opened my eyes to the fact that other mothers cared also for their children and very much more than I thought.

About a mile away there was a large field that had grown up in high black berry briars and every August people would come to gather black berries. Mother, brother Jos. B. and I went one very hot sultry day, one of those days when heavy thunder gusts are likely to come in all their terribleness scaring, as Bryant says: "______________"

Seeing that a very heavy gust was close upon us we made our way to the nearest farm house. I had known the people and had for some reason or other formed the impression that this family was a little coarse grained, and of a low order of human beings. Whether or not I was right I soon found out that that mother had very great love for her children.

She had a large family and several of the smaller children had gone with the older boys into the field where they were plowing. They had not started home soon enough and were caught in the storm. It lightened (sic) and thundered fearfully and the rain poured down in torrents. Before these children got home, that mother wept and wrung her hands and bemoaned the fate of her children, whom the said would certainly be drowned. Finally they came and it seems to me I can still se them spit the water from their mouths as that mother ran out to the road and lifted them from the horses. Ever afterwards I concluded that all mothers had great love for their children. But after all I cannot believe that my mother's love for her parents, children, and friends was certainly as strong as it could possibly be in any mother.

I recall her terrible anguish when her baby, David B., died. Then came the death of her mother and some years afterwards of her father. I think she never was quite so strong and well after David's death, and she certainly never got thru grieving after him. Mother was seldom ever sick, while father was always more or less afflicted. it had never entered my mind that she would die first but the loss of her boy David and her parents seemed to weaken her so much that when disease came she was unable to withstand its power. I was always more or less a favorite son and I know I loved my parents very much.

I cannot but recall father's anguish at mother's death. The next day he called us boys together out in the lot back of the house for I think all of us thought it too lonely to stay in the house, and he wept as I never saw him weep. Among other things he said, "O, what shall I do, what shall I do?" We as boys were also too sad and broken up to offer any suggestions.

Soon after mother's death I went to Everett, Pa., to take some lessons on the organ, for I had made up a class in instrumental music in the neighborhood where I had just been teaching public school. Somehow I felt that my teacher was not advancing me so fast as I wanted him to and I did not stay very long. I soon went to teaching and had a good class which I taught during the months of May, June and July.

As this was practically my first real teaching in instrumental music and the pupils being much scattered, I suppose I allowed myself to be somewhat easily discouraged and instead (sic) of continuing my teaching as I had expected to, and knowing that I was to take up the musical classes in the Virginia Normal school that fall, I felt that it would be good to attend the Normal to be held during August at Dayton, Virginia. This I did and was again made happy to receive first prize for my musical composition which was "Shall We All Meet at Home in the Morning" or as the tune is called, "Gathering Home." This piece has been sung quite a great deal and was my first real success. The first book this appeared in was "Good Tidings" by A.J. Showalter and I was one of the associates. This was Mr. Showalter's first really successful book and has sold into the thousands.

In this book also appeared another one of my pieces entitled "Little Ones Like Me," which is still a great favorite.

But I have often wondered why my song "Jesus is a Friend Indeed" never became a favorite for I think I rather preferred this to the others. But why one piece strikes public fancy and another does not is a question that will probably never be answered.

After this Normal I went to Bro. Jos. Click's, "My Virginia Home," as I called it and remained with them till the Va. Normal was to open. I again did some teaching in this family.

This was the first year for the school at Bridgewater and as the building (F.S. Sanger's store) over which the school could be conducted was not yet ready, the school was to be held for a short while in the public school building. The school was to have opened on the ___ of Sept. but heavy rains set in and my diary says that the school did not begin till the ___ of Sept.

I must have allowed myself to be somewhat discouraged for by what I find recorded in my diary I certainly must have been very "blue" as we often say. When we did open there were only ___ present and by Sept. ____ there were ____ in attendance.

My diary shows that I was very much undecided as to

whether I would remain at Bridgewater, or seek other fields. I think it is due to say of my special friend, Bro. Jas. R. Shipman, that it was thru his urgent and kind solicitation that I remained. But for this fact the course of my life might again here have been greatly different.

Finally the room over the Sanger building was ready to move into, but before this time I had taken the English branches from Bro. D. Hays, who was booked as one of the teachers.

After I had decided to stay I had my organ and some of my books shipped from pa. But they were long coming, for instead of their coming to Harrisonburg, Va., they went to Harrisburg, Pa.

During the 16 years that I taught in Bridgewater College I did not keep a diary regularly, for the work was routine or monotonous and I got careless.

If I can recall some of the more important events I will note same.

Index

I

J

K

S

T

U

W

Made in the USA
Monee, IL
08 November 2026

67379962R00049